MW00477432

RESTAURANT MANAGER'S POCKET HANDBOOK

MENU PRICING

25 KEYS TO

Profitable Success

DAVID V. PAVESIC, F.M.P.

Copyright© 1999 by David V. Pavesic, Ph.D.
Lebhar-Friedman Books

Lebhar-Friedman Books is a company of Lebhar-Friedman Inc.

Printed in the United States of America

Library of Congress Cataloging-in-Publication Data

Pavesic, David V.
    Restaurant manager's pocket handbook : 25 keys to
    profitable success. Menu pricing / David V. Pavesic.
        p.   cm.
    Includes index.
    ISBN 0-86730-752-8 (pbk.)
    1. Food service management--Prices.   2. Restaurant
    management--Prices.   I. Title.
    TX911.3.P7P38   1998
    647.95'0681--dc21   98-39304
                                                              CIP

PRICING is a far more complicated process than simply marking up cost. It cannot be reduced to a quantitative formula and still be effective. While such factors as food-cost percentage and gross-profit return are important residuals, the pricing process is more subjective and enigmatic than most people realize.

If you recall your last visit to a flea market, an antique mall, or a garage sale as either a potential customer or a vendor, you have experienced the uncertainty of pricing. What it all comes down to is that nothing is worth more than the price someone else is willing to pay. If you ever have sold your home or car, you probably found that what you thought you should receive and what you ended up taking were not the same.

It has been said that the buyers, not the seller, ultimately determines the price. As buyers we are not swayed by the seller telling us about his costs when we know we can get either the same or a similar product elsewhere for a lower price. Customers eating in your restaurant don't care about your costs, either. When I gave a talk several years ago at the International Hotel/Motel & Restaurant Show in New York, I began with several comments about the ridiculously high restaurant prices in Manhattan. Afterwards, one of the audience told me that I was too tough on New York operators. He maintained that they have higher overhead and, as a result, must charge higher tabs. Operators in the Big Apple have to cover costs and make a profit, or it wouldn't be worth the risk, investment, and mental anguish of running the business, he explained.

But prices must be competitive and reasonable, too. If customers don't feel that they're getting value for their money, regardless of the price, they won't make another purchase. The pricing continuum ranges from the lowest price you can charge and still make a reasonable profit to the highest price the market will bear. Most of the time, though, prices fall somewhere between those two extremes. The challenge of menu pricing is to determine where to set the price on the continuum. In addition, every item must be priced separately, and factors other than cost can impact price.

This book explores the many additional factors that must be taken into account when you price out a menu. It also covers techniques that can soften the negative impact of a price increase. Underpricing an item is as much of a problem as overpricing. One sacrifices profits, while the other may trigger complaints and a decrease in demand. Finally, the discussion of the consequences of discounting prices most certainly will change your perspective of coupon promotions.

25 KEYS TO
MENU PRICING

1... Although calculating the cost of a menu item is objective and logical, determining the price is subjective and enigmatic. **PAGE 9**

2... Pricing uses a combination of factors that involve both financial and competitive elements. **PAGE 11**

3... No single pricing method can be used on every menu item to arrive at the optimum price. **PAGE 13**

4... You cannot mark up every item by the same set amount. **PAGE 17**

5... Consider the indirect costs that impact the pricing decision. **PAGE 21**

6... Consider the important psychological aspects that enter the pricing decision. **PAGE 25**

7... Prices are either market driven or demand driven. **PAGE 29**

8... The customer ultimately will determine the price you can charge. **PAGE 33**

9... You must be able to make a profit selling an item at the price the customer is willing to pay. **PAGE 35**

10... Quality costs money, and customers are willing to pay for it. **PAGE 39**

11... Food cost is not the only factor in the pricing/markup decision. **PAGE 41**

12... You can charge more for signature items on the menu; you have a monopoly position. **PAGE 45**

Although calculating the cost of a menu item is objective and logical, determining the price is subjective and enigmatic

OF ALL THE BUSINESS DECISIONS a restaurateur must make once an operation is up and running, pricing the menu can be among the most difficult. Rational pricing methodologies traditionally have employed quantitative factors to mark up food and beverage costs. But what may seem at first to be a quantitatively based process actually requires consideration of a number of subjective factors, turning the pricing process into more of an art than a science. If pricing were a simple markup over cost, any calculator or computer program could set menu prices. Logically, price must cover your cost and return a profit.

There is a tendency to rationalize price as a means of returning an amount that will reflect a fair profit for the time, effort, and risk involved. Costs serve only as a starting point for the pricing decision. In order to remain in business, you must recover your costs. So if an item costs $3 to prepare and serve, your selling price must cover that cost.

The task of menu pricing is marked by doubts and uncertainties. Pricing decisions will determine, in part, the clientele and the amount of business the restaurant will generate — charging too much for a product or service will discourage purchases, while charging too little will reduce profit. The consumer sometimes sees prices charged as an indication of the quality of the product and service. A high price tends to imply quality, while a low price may be inferred, not as a bargain, but as low quality.

Business people are more comfortable using logical and objective criteria in the pricing decision, which is why we always want to begin by determining our cost to produce a product or service. While the cost process is objective and absolute, the pricing decision is not. Many indirect cost factors influence the price you can charge and how much the customer is willing to pay.

Pricing uses a combination of factors that involve both financial and competitive elements

THE PRICING APPROACH one uses will be influenced by each specific menu item, the existing market conditions, and the operational concept of the restaurant. For example, take the pricing of a 3-pound Dungeness crab, accompanied by fresh broccoli, red potatoes, bread, and salad, or a quarter-pound hamburger, small French fries, and medium drink. The type of restaurant, location, service style, competition, and ambiance all would impact the price of both items. The same crab dinner would be priced differently at a seafood restaurant in Atlanta than it would at a similar operation in Boston. Moreover, a quarter-pound hamburger offered at a Wendy's outlet in Columbus, Ohio, would not be priced the same as a similar burger sold in a college cafeteria in Tallahassee, Florida.

Any attempt to apply a standard markup indiscriminately to a given menu item is ineffective and dangerous because it lacks important qualitative factors that enter into the pricing decision. Qualitative considerations cannot easily be assigned quantitative values; they are subjective and require the "personal touch." And since, by definition, the personal touch tends to be biased and emotional, the task of programming those qualitative elements into a computer becomes extremely difficult.

Menu-pricing philosophies are as disparate as political and theological beliefs and almost as controversial. Initially, only raw food costs were marked up when one was setting prices. Today labor costs and even paper — in fast-food and carry-out operations — are added to raw food cost before a markup is taken. However, the financial requirements and profit demands of a business may not be compatible or realistic given existing economic or market conditions. Your cost has only a limited part to pay in the pricing decision, and customers don't care about your costs. All they're concerned about is the price they're paying. If your prices are perceived as too high relative to what the competition is charging, your quality too low, or your portions so small as to negate any price-value relationship, the customer will go elsewhere and your sales and profit objectives will go unrealized.

*No single pricing method
can be used on every
menu item to arrive at
the optimum price*

IF PRICING WERE SIMPLY a quantitative function of cost markup, we could program a computer to price out every menu item. Our objective is to recover our costs and realize a fair profit for our time, investment, and risk. Most pricing methodologies start by calculating the cost of the menu item. We note what the various ingredients cost and calculate our cost per servable portion based on yields. With trim loss or cooking shrinkage, the cost per servable pound always will be greater than our cost per pound as purchased.

Everybody has his or her favorite method of arriving at the price to charge. Some methods are highly complicated and attempt to take into account everything from food cost to labor and overhead. Others are relatively simple, such as marking up an item three times cost

> **"A man who never makes mistakes loses a great many chances to learn something."**
>
> — ANONYMOUS

plus a dollar. You can find a number of pricing systems covered in articles and books written on the subject, which are all useful to a point. But whatever method you finally elect to use, you will not be able to use it to price every item on the menu.

A number of methods ranging from multiplier factors to complicated formulas can be employed to mark up cost. Some attempt to guarantee the return of a minimum profit, while others return a desired food-cost percentage or gross-profit amount. How one arrives at the multiplier factors, food-cost percentage, or gross-profit amount is more important and relevant than the actual mathematical calculations. Even if one were to assume that the basis for the pricing strategy was sound and practical, it could not be applied to every item on the menu without overpricing or underpricing most items.

Certainly, a formula can be helpful by providing a starting point from which to adjust your price upward or downward, depending on the competition, your restaurant's position in the market, or the price points your operation seeks to offer. But remember: Pricing cannot be driven by some formula. Moreover, all items cannot be marked up the same amount. Pricing or markup is almost never exclusively cost-driven.

A typical menu comprises items with varying food and labor costs. For instance, soft drinks have a low food and labor cost, while items like steaks and seafood have high food costs and, in some cases, require considerable labor to prepare. Consequently, items with low costs will be priced lower than items with high costs. That results in some items being underpriced while others are overpriced.

Appetizer and dessert prices cannot exceed the prices of the entrées. The entrée price has the greatest influence on the average check per customer. If appetizers and desserts are priced more than half the price of an average entrée, customers will be disinclined to purchase them along with a main dish. What often happens in such a case is that a customer will order an appetizer as an entrée or split it with another person.

Proper pricing cannot be accomplished by forcing the price to achieve a particular food cost, gross profit, or average check. Each menu category — appetizers, salads, sandwiches, entrées, and desserts — must be marked up independently and differently to arrive at the price the customer is willing to pay.

(1) All items on a menu should be marked up the same amount.

    A. True
    B. False

(2) Appetizers and desserts should be priced:

    A. More than half the price of an average entrée
    B. At whatever cost the customer will pay
    C. In balance with all menu items

(3) Serving soft drinks involves:

    A. High food and high labor costs
    B. Low food and low labor costs
    C. Low food and high labor costs

(4) An example of items added to a raw food cost can include:

    A. Paper
    B. Labor
    C. Soap

(5) Lobster should be priced the same in Boston and Des Moines.

    A. True
    B. False

ANSWERS: 1: B, 2: C, 3: B, 4: A, 5: B

You cannot mark up every menu item by the same set amount

IF EVERY ITEM ON THE MENU were marked up to achieve the same food-cost percentage, you wouldn't need to conduct menu-sales analysis. In fact, back in the 1960s, one successful family restaurant chain experimented with marking up every item the same amount but then abandoned that methodology very quickly. The trouble with viewing the pricing decision as a quantitative exercise of cost markup is that while cost is important, it is only one part of the package used to arrive at a price.

Applying a single markup method or marking up each item to return the same profit will result in items being both overpriced and underpriced compared with what is customary in your market. If, for example, your cost to produce a common menu item, such as a hamburger, is twice the cost of other restaurants,

"Diamonds are nothing more than chunks of coal that stuck to their jobs."

— MALCOLM FORBES

your price cannot be twice that of your competition. Your cost is excessive, and customers won't pay your price. You can charge only the customary market price for a burger unless you turned it into a specialty item. Markup is tempered by other factors besides food costs and desired profit.

Remember: Pricing is not an exclusively cost-driven process. Using a cup of beef bouillon as an example, the food cost of a 5-ounce cup would be about 10 cents. If it were priced to achieve a 40-percent food cost (a markup of two-and-a-half times its raw food cost), the bouillon would be underpriced at 25 cents. Since the item usually is served in white-tablecloth operations, the customary price point could be $3.95, and it still wouldn't be considered overpriced by patrons. One can charge more for ambiance, customer service, amenities, and other indirect cost factors.

In the same restaurant, pricing a 1.5-pound Maine lobster that costs $6.95 per pound would require a menu price of $26.00 without adding such accompaniments as salad, bread and butter, and vegetable. That would appear to be within an acceptable price range, given the menu item and the type of restaurant. Consider that when a single markup is applied to every item on the menu, the assumption is being made that costs are distributed equally. While overhead costs are allocated evenly to all menu items, certain direct costs of producing a particular dish should be charged to only those items incurring that direct cost — such as a meat cutter, baker, and items prepared from scratch. Other items are purchased prepared and pre-cut, requiring only final heating for service. The pricing of such items must take those things into account, and convenience items do not require as high a markup as others that need processing and handling.

Each menu category must be priced differently, thereby making it impractical for a single markup method to be applied. Whenever a single method is employed, it makes several incorrect assumptions. The first is that pricing is not exclusively cost driven. Other indirect factors must be considered in order to arrive at the price to charge. Second, it assumes that costs are divided equally across all menu items. Some items demand little labor to produce, while others take up much time and require skilled culinarians to prepare. A single markup method charges the same amount of labor and overhead to every item.

Third, as previously mentioned, a single markup method will overprice high-cost items and underprice low-cost items. And finally, a single pricing methodology fails to consider what I refer to as the "price-volume factor."

What that means is that popular items do not have to be marked up as much as slower items to return the desired profit or food cost. Price does influence popularity, and if you price something too high, it will not sell; conversely, if you price an item too low, you lose potential profit. Therefore, the pricing decision requires that you use multiple measures to price out your menu.

Consider the indirect factors that impact the pricing decision

INDIRECT COSTS are those that cannot be assigned to specific menu items but are factors that customers recognize as the value-added aspects of having chosen your restaurant over that of a competitor. For example, if your operation is considered the top steak house or Italian restaurant in your market, you will be able to charge more than the customary menu prices. Similarly, if your service is head and shoulders above that of most of your competition, you can charge more than the average operation.

In other words, some things that you do give you a competitive edge; consequently, you can charge a little more. Examples of other indirect cost factors that allow you to adjust prices upward (or downward, if they are absent) are ambiance, location, amenities, product presen-

"You are only doing your best when you are trying to improve on what you are doing."

— ANONYMOUS

---

tation, customer demographics, and specialty menu items. You can make further adjustments to fine-tune menu prices that involve your desired check averages and the high and low price points of your various menu-item categories.

Let me provide some examples of what I mean by those indirect cost factors. If your restaurant is considered the premier operation of its type in the market, you can price more aggressively than if you were just marginally competitive and a follower rather than a leader. The atmosphere and decor of a restaurant adds much to the enjoyment of any meal, and the perception of value is enhanced when the restaurant is decorated tastefully. Or perhaps it is the service commitment that makes your restaurant competitively distinctive. Those intangibles are recognized as added values by the customer and can be reflected in your prices.

The pricing structure can dictate the status or economic class of the clientele a restaurant will attract. The location of the restaurant can have a lot to say about who your customers are. If, for example, you're located in an area of business offices and high-rise condominiums, and the inhabitants of those buildings are your target clientele, your prices can be higher than if you were located in the suburbs and your customer base were middle-income families.

The aspect of product presentation proclaims, "Sell the sizzle, not the steak." Product presentation also is important in forming the value perception of the customer. A beautiful plate presentation and custom-designed china and glassware really can enhance the appearance of the food or alcoholic beverages served — and often allow you to charge a higher price. Those are just a few of the subjective factors that should be considered in the pricing decision.

(1) "Sell the sizzle, not the steak" means:

   A. Include cooking costs in pricing steaks
   B. Quality of meat should not be considered
   C. Product presentation is important in value perception

(2) Indirect cost factors include:

   A. The location of the restaurant
   B. The ambiance of the restaurant
   C. The amenities of the restaurant
   D. All of the above
   E. None of the above

(3) Indirect cost factors should not be considered in menu pricing.

   A. True
   B. False

(4) If your operation is considered the top steak house in the region:

   A. You should consider slightly lower prices
   B. You should consider slightly higher prices
   C. You should consider serving seafood

(5) If service is a problem in your restaurant, you may consider:

   A. Solving the problem but keeping prices high
   B. Solving the problem and lowering prices until the problem is resolved
   C. Firing the staff and raising prices with the labor cost surplus

ANSWERS: 1: C, 2: D, 3: B, 4: B, 5: B.

Consider the important psychological aspects that enter the pricing decision

WHEREAS THE INDIRECT COST FACTORS that influence the pricing decision are conditions controlled or established by the operator, psychological aspects of menu pricing reflect the attitudes and images in the mind of the customer. The psychological aspects of menu pricing include: odd-cents pricing, mental accounting, reference pricing, and time-and-place factors.

Odd-cents pricing is a technique employed in retail pricing for everything from clothing to real estate. Prices are stated in amounts that end in a number nine or five. Psychologically, that is viewed as a lower price than if the number ended in a zero. For example, $9.95 is preferable to $10.00. The number is three digits instead of four and often is rounded down mentally to $9.00 instead of up to $10.00. The element of low-price perception is an important

> **"When you're in business, there is no such thing as making money. If you made $50,000 and could have made $100,000, you did not make $50,000 — you lost $50,000."**
>
> — ANONYMOUS

pricing strategy. I recommend that you set your menu prices with digits ending in .25, .50, .75, and .95. And when incremental increases force you to move up to the next highest price, the customer rarely detects the hike.

Odd-cents pricing tends to reduce the price stress; the customer goes through a process of "mental accounting" when making a purchase. The theory suggests that customers will sort food purchases into budget categories of either groceries, entertainment, or social expenditures. Each budget category is controlled to some degree by a spending restraint. Consequently, the amount spent on a meal purchased away from home will vary depending on whether the expenditure is debited to the grocery- or entertainment-expense categories.

Generally speaking, a person spends more freely when the entertainment or social account is being debited. If a person eats out during the week rather than cooking at home, money usually is debited to the grocery budget, not the entertainment budget, and the expenditure tends to be influenced by price and convenience. As a result, the expenditure will be considerably less than if the person had dined out on the weekend for social or recreational purposes. With that knowledge you can price your menu items more effectively.

As an operator, you must determine how the customer categorizes your restaurant — that is, eat-out or dine-out. For example, a meal at a fast-food restaurant is likely to be considered as a food/grocery expenditure. Consequently, if your restaurant wants to attract that customer during the week, nightly specials would need to be offered and priced accordingly. Have you noticed that most casual-theme restaurants do not honor discounts on Friday and Saturday nights? The reason is that people are "dining-out," and budget restraints are relaxed.

If we follow the logic of that mental accounting theory, the pricing decision can be approached from the consumer's perspective. The objective is to shift the expenditure into a higher budget category or combine it with another category. The budget category can change, depending upon the occasion, such as a birthday, anniversary, or even the day of the week. To help entice weeknight customers on grocery budgets to eat out instead of at home, an operator may offer promotions like early-bird specials and discount coupons.

Most customers do some type of price comparison when they shop for shoes and clothing, so it shouldn't be surprising that they would do the same when eating out. When a competitor opens in your market, you should check it out by comparing comparable menu prices with yours. Customers do the same thing. If the prices of the new restaurant are lower than those in their "reference" restaurant, they perceive a good price-value. But if they're higher than the reference price, the price-value is diminished. The more one pays, the more critical he or she will be of the food and service. If you charge at the high end of the market for comparable food and service, expect customers to be more critical of even the smallest details.

The last psychological aspect of menu pricing has to do with where and when you make your purchase. That is called time and place. It's best explained in the context of your price tolerance for a hamburger and a Coke at a sports stadium or theme park compared with the price you would be willing to pay at your neighborhood restaurant. Another example is your price tolerance when on vacation versus eating out on a weeknight while you are at home. Such knowledge is needed to determine effectively which items to offer and how they should be priced to appeal to your customers and optimize your revenue.

Prices are either market driven or demand driven

A QUESTION OFTEN PONDERED by restaurateurs is, "What price should I charge?" Two extreme pricing philosophies define the limits of the pricing continuum. The pricing continuum ranges from *the highest price the market will bear* to *the lowest price you can charge and still return a fair profit*. The price you charge for a given menu item will fall somewhere on that continuum. The question is whether it will fall on the high side or the low side. Pricing can be either *market driven* or *demand driven*, and the approach one adopts will depend upon the menu items and the operational concept. If the menu item is a "commodity" in the economic sense — that is, it is available just about everywhere and quality differences are nominal — a definite price point exists in the market and in the mind of the customer.

> **"Identify with ownership interests. It is the fastest way to be recognized by your boss."**
>
> — ANONYMOUS

Some examples of commodity menu items are your basic pepperoni pizza, a quarter-pound hamburger, and barbecued ribs. If there is nothing special about the taste, preparation methods, or presentation, the price charged must fall in line with the prices of your competitors. That perspective also is applied when you introduce new menu items before any substantial demand has been established. Prices that are market driven tend to be on the moderate-to-low side of the pricing continuum.

Demand-driven prices can be used for items that are offered by only a few competitors. Overall pricing can be demand driven when you have customers waiting in line every night to eat in your restaurant. You can be aggressive in your pricing and be on the high side of the pricing continuum. However, if you charge the highest price your clientele will pay, expect customers to be more critical of food quality and service. You also will have to be on the leading edge of food trends and quality, decor, and service. Remember: Your customers are paying for it!

The opposite end of the pricing continuum is *the lowest price that will still return a fair profit*. That is an approach taken by those wishing to appeal to the largest share of customer market. Offering large portions and high-quality ingredients will win you customers and compliments. This pricing strategy is employed in highly competitive markets. However, operations with low prices and high customer counts may not be optimizing their profit potential on each sale. With long waiting lines for tables, it would appear that they could increase prices on some items without significantly impacting demand.

One thing to always keep in mind is that no restaurant will be able to sustain a product and price advantage over the long run. Competition eventually will add those items to their menus, changing a demand-driven price to a market-driven price.

(1) Demand-driven prices can be used for:
   A. All items on a menu
   B. Desserts only
   C. Items only offered by a few competitors

(2) If you charge the highest price your clientele will pay, then:
   A. Expect customers to charge meals with credit cards
   B. Expect customers to wait in line
   C. Expect customers to be more critical of food and service

(3) A restaurant can sustain a product and price advantage indefinitely.
   A. True
   B. False

(4) Menu pricing can be:
   A. Both market driven and demand driven
   B. Either market driven or demand driven
   C. Neither market driven nor demand driven

(5) Customers are more price tolerant:
   A. While eating out on a weeknight at a family restaurant
   B. While preoccupied with conversation when ordering
   C. While on vacation

ANSWERS: 1: C, 2: C, 3: B, 4: B, 5: C

The customer ultimately will determine the optimum price you can charge

TODAY WE HAVE A PLETHORA of software programs available for managing food costs. Recipes are costed automatically and updated as often as delivery invoices are recorded. Variance reports alert management to excessive inventories and ingredient usage based on what is selling. Having all of that technology at our fingertips seems to imply that pricing decisions also can be monitored and controlled. But while the accuracy of food costing has reached a new level of sophistication as a result of incorporating point-of-sale systems with menu and inventory-analysis capabilities, the pricing decision remains one of subjectivity and guesswork.

Pricing a menu item is more of an art than a science. Although cost is important to us as owner-operators, it is not viewed that way by our customers. When a customer points out that the price of an entrée is higher than your competitor's, explaining that your costs are greater and therefore you must charge more won't make him feel any better.

Put the shoe on the other foot for a moment. If you have traveled on your own dollar to, say, the National Restaurant Show in Chicago, you probably called around searching for the cheapest airfare. If you waited too long to make reservations, you paid more than someone who booked at least 14 days in advance. On the other hand, if you flew on a discount airline instead of a major carrier, you probably saved on the ticket. But whether you paid $200 or $400 for the seat, you arrived at your destination just the same. Consider your feelings about the price you paid in each scenario.

As a consumer, if I know I can get the same or comparable accommodations, transportation, or meal at a better price by shopping and comparing, I will do just that. If costs are driving up your prices to the point that customers are complaining, you must find a way to reduce costs. You cannot be like the post office, which just keeps raising the price of first-class postage to cover cost increases. In a competitive marketplace, customers will vote with their legs and move on to the operation they believe will give them value for their dollar. Therefore, the challenge to you as an operator is to be able to make your profit selling at prices the customer is willing to pay. Your costs are not relevant to them.

You must be able to make a profit selling an item at the price the customer is willing to pay

THAT STATEMENT IMPLIES that there is a maximum amount you can charge for any given item on your menu. Now while you want to price at, or as close to, the top limit of that price threshold, your costs must allow you make your profit selling at that price. That applies to every item on the menu, from the ubiquitous hamburger to the specialty items on which you built your restaurant's reputation.

While what is considered "expensive" is relative to income and spending habits, restaurants and hotels in any metropolitan area also are evaluated by what we are accustomed to paying. Whether we're in New York, Atlanta, Chicago, or Los Angles, we will always compare prices with those back home. What is considered expensive in Tallahassee may be a bargain-basement price to someone from Miami or

Boston. When tourists from the Northeast dined at my Italian restaurant in Orlando, Fla., they often would marvel at our prices. They would claim that they could feed their entire family there for what it cost for two people back home — plus the portions were larger, they added.

In that case our customers expected to pay more than they were charged; consequently, they left happy, feeling that they had received value for their dollar. However, there will also be the customer who feels that your prices are way out of line with what he expected to pay. Again, it is all relative to one's social and income level and his or her frequency of dining out.

Pricing also can vary within a particular area. In Atlanta, for example, the price point for lunch is about $6 with tax, tip, and beverage. If you expect to get people to spend money in your restaurant on a regular basis, you must offer them several choices around that price. The higher your overhead and labor cost, the more difficult it is to provide viable menu selections at that price point and still make a fair profit. Your operating costs will impact the prices you must charge to cover direct costs and overhead and return a profit. If your overhead is too high, you may not be able to price competitively. While covering costs is imperative to maintaining solvency, if costs are not contained or are excessive, you cannot maintain your market share by raising prices to cover your cost inefficiencies.

Therefore, it is imperative that you know who your customer is and price according to his or her expectations. If your restaurant draws customers from a retirement community, their expectations will be different from those customers who live in an upper-middle-income residential area. You also must understand how your restaurant is perceived by your customers. Whether they view you as an "eat-out" or "dine-out" operation will impact the prices they are willing to pay. If your customers are on a limited budget for eating out during the week, they will look for low-price specials. Consequently, you must base your prices on what the customer is willing to pay and develop menu items that can return the profit you require to remain in business. Profits cannot be maintained by discounting items initially priced to return your desired profit.

(1) There is a maximum amount you can charge for any given menu item.
- A. True
- B. False

(2) What is considered "expensive" is relative to:
- A. Your labor cost
- B. Your food cost
- C. Income and spending habits in the region

(3) An example of a competitive market is:
- A. Several chain restaurants in a shopping mall outside Chicago
- B. The postal service in the United States
- C. The one and only Mexican restaurant in Dubuque, Iowa

(4) Menu pricing involves knowing who your customer is, and:
- A. Pricing according to the labor cost formulas
- B. Pricing according to his or her ethnic background
- C. Pricing according to his or her expectations

(5) The following will most likely impact prices your customers will pay:
- A. The nearest residential area is upper-middle-income
- B. The nearest gas station is 3 miles away
- C. The nearest theater is 3 miles away

ANSWERS: 1: A, 2: C, 3: A, 4: C, 5: A.

KEY

Quality costs money, and customers are willing to pay for it

IF YOU PROVIDE QUALITY food and service, you can charge more than those who do not. While there are certainly price-points that must be considered when you are setting menu prices, the public has demonstrated that it's willing to pay a premium for quality. That's true regardless of which end of the pricing continuum your prices fall on or the segment of the market you serve.

If you're a low-price, fast-food operation that uses only freshly ground beef, top-quality brand condiments, and a premium french-fry brand, you can charge a little more than those who use the least-expensive products. You don't control food cost by purchasing the cheapest ingredients. I have a favorite quote about quality and prices that I first saw hanging in a Baskin-Robbins ice cream shop many years ago. It was by British writer and reformer John Ruskin, who said, "There is always someone willing to make something a little worse and sell it a little cheaper; and those who consider price only are this man's lawful prey." I can just picture a price-conscious customer complaining to Mr. Baskin or Mr. Robbins about having to pay 15 cents a scoop for their premium-quality ice cream when the going price per scoop for ordinary ice cream was still 10 cents.

There is chocolate, and there is Godiva; there are hotels, and there is the Ritz-Carlton; there are foreign luxury cars, and there is Mercedes; there are Swiss watches, and there is Rolex. While a market certainly exists for low-end chocolate, hotel rooms, foreign cars, and watches, the point is that no matter where your product or service falls on the pricing continuum, quality is recognized by the customer, and it's a competitive advantage if your prices are perceived as fair. With the proper cost controls, you are able to offer higher quality and still price competitively to achieve a decent return for your time and investment.

11

Food cost is not the only factor in the pricing/markup decision

THE MOST FREQUENTLY USED pricing and markup methods are built around achieving a target food-cost percentage. While that's a logical place to start the pricing process, keep in mind that it's just one of the cost components. Pricing every item to achieve a standard food-cost percentage may at first seem like a good idea. But you will realize quickly that this pricing method will overprice high food-cost items and underprice low food-cost items. Items like soft drinks, pasta dishes, some chicken items, soups, and salads have low food costs, but because of their popularity with customers, they can be marked up higher than average. Meanwhile, such items as steaks and seafood are marked up less to achieve competitive price points.

Some menu items require more direct labor to produce than others. In fact, cafeteria

> **"Demand quality and be willing to pay for it; however, do not assume that higher prices always mean higher quality."**

> — ANONYMOUS

owner-operator Harry Pope was the first to note that the labor cost of preparing certain menu items can be higher than the food cost itself. Therefore, marking up the food-cost portion isn't enough. To accommodate added costs, Pope included the cost of the *direct labor* along with the raw food cost of the menu item and then applied his markup. The inclusion of direct labor with food cost is called *prime cost*. Pope divided his menu items into two categories: items requiring little or no direct labor and items requiring direct labor. And he priced each differently.

Any attempt to mark up every item by the same amount also fails to consider what I call the "price-volume relationships." What that means is that you can mark up popular items less than slow-sellers because the total gross profit realized from selling 30 or 40 orders per night compensates for the lower markup. The old retail saw that says, "Volume allows us to sell at lower prices than our competition," applies to menu items as well. When it comes to pricing slow-selling items — especially those with expensive, perishable ingredients — one feels justified applying a higher markup because of the risk of loss. However, a higher markup may be counterproductive if it discourages purchase. A more moderate markup may encourage purchase and frequent turnover of perishable and expensive ingredients.

The menu category — appetizers, entrées, desserts — also influences the markup and price you can charge. You won't sell many appetizers or desserts that are priced within a dollar or two of the average entrée. So price spreads between menu categories is another variable in the pricing decision. Pricing will never be a formula-driven process. There are too many subjective variables that cannot be quantified in a formula.

(1) Marking up every menu item by the same amount fails to consider:

    A. The time period lunch is served
    B. The price-volume relationships
    C. The labor-supply relationships

(2) You won't sell many appetizers if:

    A. They are priced within a dollar or two of the average entrée
    B. They are priced competitively with other restaurants in the area
    C. They are priced at half the cost of an average entrée

(3) In general, you can mark up popular items less than slow-sellers.

    A. True
    B. False

(4) A "prime cost" is defined as:

    A. The cost of prime-grade meats
    B. The cost of serving menu items during prime hours
    C. The inclusion of direct labor with food cost

(5) Menu pricing is solely a formula-driven process.

    A. True
    B. False

ANSWERS: 1: B, 2: A, 3: A, 4: C, 5: B

You can charge more for signature items on the menu; you have a monopoly position

EACH MENU ITEM MUST BE PRICED on its own. The unique preparation method and plate presentation may allow you to mark it up much higher and return an even lower food-cost percentage. Each restaurant should have several house specialties or signature items — menu items that are available only at your restaurant or prepared or served in your inimitable style. Customers may visit your restaurant specifically to order your linguine with white clam sauce, for example. Or the lines could be forming for your barbecued ribs; honey-basted, mesquite-broiled half chicken; or to-die-for homemade chicken and dumplings.

If you're known for a particular menu item, you can charge above the average for that item because it is, in an economic sense, a "specialty good." Ordinary barbecued ribs or plain broiled chicken are commodities or generic menu items without any real identity. Store-brand milk and bread are commodities in our supermarkets and are priced competitively. The name-brand breads and milk products are similar in terms of ingredients, yet are priced 10 percent to 20 percent higher.

If you have a monopoly on one or more menu items, you can price them near the higher end of the pricing continuum. When you sell out every night or have a line waiting to be seated every night, test for price elasticity on certain popular items. Just an additional 20 cents or 25 cents can lower costs and increase gross profit without appreciably lowering demand or customer price-value perceptions.

For discount pricing to be economically beneficial, customer counts must increase significantly in order to offset the reduction in price

YOU MIGHT GET THE IMPRESSION from the number of discount promotions being offered that discounting is a simple and effective way to increase business. The use of discount coupons is so pervasive that regular prices are meaningless. Discounts are renewed month after month so that customers come to expect some kind of discount on every purchase. Well, I'm going on the record to tell you that discounting doesn't make financial success unless traffic and sales increase.

If all you have going for you is your discounted price, you will not sustain your market share for long. If couponing attracted new customers or helped you introduce a new product or service, that would be a positive result. But the truth is, that regular customers — those who have been frequenting your operation and

> **"An executive is a person who always decides; sometimes he decides correctly, but he always decides."**
>
> — JOHN H. PATTERSON

---

paying full menu prices — will begin to use the discount coupons. The only way you can show a gain in that situation is if regular customers increase their frequency of visits during the week.

Many restaurant operators remain unconvinced that discounting is always a cost-effective pricing strategy. To me, it's like leading with your chin in a boxing match; you're positioning yourself to get hit. If you believe that low price is the only consideration customers use when deciding where to eat, fine-dining operations would never survive. In fact, customers who consider only price are usually not loyal customers to any particular operation. Their loyalty lasts only as long as the discount is in effect; then they switch to another restaurant willing to "purchase" their loyalty with a price discount.

If you're planning to enter the coupon wars, keep a few things in mind. You must consider the consequences inherent in discounting when you determine the actual percentage of price reduction you will offer. You must understand fully the impact that discounting has on your overall financial results.

Whenever you offer a discount, profit margins will be lowered. In some cases the deep discount may barely cover your costs. If lowering your price brings in more customers, demand for your business is elastic or price sensitive. Failure to realize sufficient increases in sales volume to offset the lower margins of each discounted sale can have an adverse effect on financial outcomes.

Don't engage in discounting if you're operating below the break-even point. It will take you longer to reach a break-even sales point because of the reduced margins on each discounted purchase. In addition, the higher the discount value — or the greater the percentage of discounted transactions — the greater the need for increased traffic to achieve the same bottom-line profit as before the discount.

Sales volume and customer counts must improve simply to offset the cost of discounting. Since discounting reduces the profit margin on each transaction — consider the impact of a buy-one-get-one-free promo — the break-even point increases because the food-cost percentage rises. You're using food for two orders and collecting money for only one. That variable cost increase can be offset only by a decrease in fixed-cost percentages, which can only occur through increased sales.

(1) Don't engage in discounting if:
- A. Your operation is above the break-even point
- B. Your operation is below the break-even point
- C. Your operation is involved in coupon wars

(2) Discounting reduces the following in your operation:
- A. The profit margin
- B. The hours of operation
- C. The number of customers

(3) Customers who consider price are not usually loyal customers.
- A. True
- B. False

(4) If your restaurant is known for a particular menu item, you can:
- A. Be less concerned with service
- B. Charge less for that item
- C. Charge more for that item

(5) If you offer a buy-one-get-one-free promotion, your break even point:
- A. Decreases
- B. Stays the same
- C. Increases

ANSWERS: 1: B, 2: A, 3: A, 4: C, 5: C

14

You can calculate the additional customer traffic needed to offset the reduced margins of the price discounts

EACH DISCOUNT PROMOTION has what I refer to as a "trigger point" of increased volume that must be achieved to offset the reduced margins of each discounted sale. It employs a variation of the traditional break-even formula. You must first make some estimate of the redemption rate of the coupon promotion. The redemption rate will, in part, be influenced by the way the coupon is disseminated — that is, direct mail, newspaper, coupon booklet — and the amount of the discount. A "buy-one-get-one-free" will pull more than "buy-one-get-second-at-half price."

The example here assumes that the discount is the "second-at-half-price," and that 10 percent of the customers will redeem coupons. That discount is equivalent to a 5-percent reduction in sales revenue $(.10 (.50) + .90 (1.00)$

> **"The moment avoiding failure becomes your motivation, you're down the path of inactivity. You stumble only if you're moving."**
>
> — ROBERTO GOIZUETA

= .95). That says that 10 percent of your customers will receive a 50-percent discount, and 90 percent will pay full price. If pre-discount sales ran $1,000 for 100 customers, the average check is $10 per person. With the discount only $950 in sales would have been received. Assuming a 38-percent food cost, or $380, if only $950 were collected, the food cost would increase to 40 percent ($380/$950).

If prediscount variable costs totaled 54 percent, they would increase at least to 56 percent with the discount. Because sales did not increase, the fixed-cost percentage did not decrease. We can determine the increase in sales necessary to be no worse off than before the discount. The formula is:

$$\frac{1}{1-VC\% \text{ w/discount}} \text{ (increase in variable cost \% w/discount)}$$

The sales increase required to offset the 2-percent increase in variable cost for this example is:

$$\frac{1}{1-.56}\ (2)\ =\ \frac{1}{.44}\ (2)\ =\ 4.545\%$$

$.4545^1 \times \$1,000 = \$45.45^2$
$\$1,000 = \$1,045.45^3$

[1] 4.545% expressed as a decimal.
[2] Additional sales over and above $1,000 to offset discounting.
[3] Total sales needed to offset discounting.

A 2-percent increase in variable cost can be offset only by a 4.54-percent increase in sales, which will reduce fixed-cost percentage by 2 percent. That can be converted to reflect the increase in customer count by dividing the new sales figure by the average check with the discount, or $1,045.45/9.50 = 110 covers. That is 10 more customers than the average without the discount. Incremental profit would result only when the customer count exceeds 110. Remember: Even with 110 discounted covers, you are still no better off financially than you were serving 100 covers at regular prices.

(1) What percentage is a reasonable assumption for food cost?

    A. 18%
    B. 28%
    C. 38%

(2) Which promotion may draw more customers into your restaurant?

    A. Buy-one-get-one-free
    B. Buy-one-get-second-at-half-price
    C. Buy-two-get-one-free

(3) A 2-percent increase in variable cost can be offset by a _____ increase in sales.

    A. 2.54-percent
    B. 5.54-percent
    C. 4.54-percent

(4) You can calculate the additional customer traffic needed to offset the reduced margins of price discounts.

    A. True
    B. False

(5) A discount promotion usually has a "trigger point" of increased sales volume that must be met to offset the reduced margins of a discounted sale.

    A. True
    B. False

ANSWERS: 1:C, 2:A, 3:C, 4:A, 5:A.

Your average check is more than just the calculation of total sales divided by total customers served

IF ASKED TO CALCULATE your average guest check, you probably would just divide the food and beverage sales by the number of customers served. Average check is a common way to monitor sales at breakfast, lunch, and dinner. But you should realize that calculating your check average tells you what your customers currently are spending — and it's not necessarily the average you must realize to meet your daily sales goals.

The desired check average is determined before the menu is priced. If you can forecast the number of customers you will serve at each meal period, you will be able to influence your sales revenues by making it impossible for a customer to spend less than the predetermined amount needed to achieve your sales projections.

"Problems are only opportunities in work clothes."

— HENRY KAISER

Assume that your pro-forma income statement tells you that your operation will break even if you serve an average of 250 customers who each spend about $15.00 on food and beverages. You now must price and design your menu to realize a desired average check of $15.00 per person. If you calculate your average check and find that it is less than that amount, look first at your menu pricing and design as a possible reason.

If you have too many entrées priced below $11.95, you may be "encouraging" customers to purchase lower-priced entrées and will find it necessary to sell more appetizers and desserts to achieve the desired average check. If the majority of your entrées are priced between $13.95 and $15.95, you will achieve your target check average. Most customers assume that your menu items are priced competitively and that price-value is present. Forcing a price-point or check average will not be effective. Price-value relationships always must be considered. Don't try to force up prices on commonplace items, such as steaks and Prime rib. At the same time, a steak house using its steak trimmings in a chopped sirloin priced at $8.95 may appear to be offering a bargain when its regular steaks are priced up to $24.95. One must be aware of the desired check average and control the number of entrées priced above and below that target figure.

Have you ever noticed that the price of a trip to the salad bar is within a dollar or two of the lowest-priced complete dinner that includes the salad bar? That is done partially to achieve a minimum average check from each guest and to encourage the purchase of dinners rather than à la carte salads.

Therefore, regard your desired average check as you would your food cost or labor-cost percentages. It's a standard or benchmark that you expect to achieve if your sales goals are to be realized. Adjustments in pricing and menu design must be made if you are unable to achieve your standard. If your menu is designed and priced properly, it will be impossible for any customer to spend less than your average check. You cannot always rely on servers to up-sell customers and build the check average. Your printed menu, if priced and designed properly, is your most consistent and best sales tool.

QUICK QUIZ

(1) The desired check average is determined:

    A. After the menu is priced
    B. Before the menu is priced
    C. Regardless of when the menu is priced

(2) If your menu is designed and priced properly, it will be impossible for any customer:

    A. To spend less than your average check
    B. To spend more than your average check
    C. To pass up ordering desserts

(3) Usually a trip to the salad bar is priced within _____ of the lowest-priced complete dinner that includes the salad bar.

    A. Five or six dollars
    B. One or two dollars
    C. Twenty-five to thirty-five cents

(4) Price-value relationships should be considered:

    A. Only for appetizers
    B. Only for entrees
    C. For all menu-pricing decisions

(5) What is your restaurant's best sales tool?

    A. The printed menu
    B. Television advertising
    C. Word-of-mouth referrals

ANSWERS: 1:B, 2:A, 3:B, 4:C, 5:A.

16

Every restaurant is placed into a price category by its customers

CUSTOMERS WILL PLACE A RESTAURANT into one of three price categories: (1) low-priced; (2) moderate-priced; and (3) high-priced. Specific numerical check averages are not given because customers, depending upon their incomes and geographic location, will apply their own dollar ranges when placing restaurants into one of the categories. For example, the prices charged by a luxury restaurant in an Atlanta suburb would be lower than those listed by similar operations in Manhattan or Los Angeles.

In any case, the restaurant's menu prices must fall in line with the price category in which the majority of its customers place the operation. If some menu items exceed that range, customers will purchase fewer of those items. Conversely, if items are priced too low, the operation runs the risk of lowering the overall image, and the average check may drop.

Part of that categorization is influenced by whether the dining occasion is an "eat-out" or a "dine-out" experience. If a restaurant is considered an eat-out operation during the week — a substitute for cooking at home — customers will be more price-conscious. If a restaurant is categorized as a dine-out operation, the visit is regarded more as a social and entertainment occasion, and spending restraints are more relaxed.

Knowing how patrons evaluate your restaurant is important to your pricing decisions. Rarely will a restaurant be rated in both categories by the same patron. The frequency of visits to an eat-out operation will be greater than to the dine-out operation, but the amount of money spent will be considerably less. Regular weekday customers may go elsewhere for special celebrations like anniversaries and birthdays. Weekend clientele differ somewhat from midweek customers. For example, local residents may be the bulk of the traffic during the week, while visitors, tourists, or people traveling from outside the restaurant's normal market area may make up the majority of weekend business. Such patrons may categorize the operation as a dine-out, or special-occasion, restaurant and will spend more liberally.

Be aware of the price differential between similar or competing menu items

THE PRICE DIFFERENTIAL between similar or competing items on the menu is another reason for adjusting prices upward or downward after initial pricing has been completed. For example, consider the competitive similarities between baked or broiled chicken and roast duck. The cost of chicken is much lower than that of duck, but the duck returns a higher gross profit. If the price of the chicken is too close to that of the duck, customers will order more duck for pricing reasons: "For $1.50 more I can get duck instead of chicken." The pricing strategy is not necessarily to lower the price of the chicken, but to raise the price of the duck. Duck is a prestige menu item, and customers expect to pay more for a duck dinner than they would for chicken. Chicken is perceived as inexpensive, no matter how it is prepared, and therefore will have a price limit far

"The only real mistake is the one from which we learn nothing."

— JOHN POWELL

---

below that of what one could charge for duck — even if a larger portion of chicken is served. If a menu offers both, the price differential between the two could be a determining factor in the customer's selection process.

Two apparently similar veal entrées could present the same dilemma. However, one might require a considerable amount of labor to produce, while the other needs little prep time or labor. In that case the former should be priced higher so the patron's choice is based more on preparation methods and ingredients. If the price spread is more than 25 percent of the higher-priced item, the number sold of the lower-priced chicken or veal selection may increase. Conversely, if you want to sell more duck or expensive veal, the price of the lower-cost chicken or veal dishes should be increased so the differential is less than 25 percent. That will make the price of the duck and expensive veal dishes more appealing.

That distinction is being raised because customers are prepared to substitute certain items for others. Those products are called competitive or substitute items, and when a person buys one of them, he or she usually purchases less of another item. Consequently, if more duck is purchased, the demand for chicken is reduced. That should alert the menu planner to limit the number of substitute or competitive items on the menu or else use a preparation method that is significantly different for each one — fried or barbecued, for example — making it less of a substitute.

In contrast, complementary items are those whose sale boosts the sale of other products. Examples are à la carte accompaniments to entrées and desserts, such as marinated mushrooms on steaks and ice cream on hot apple pie. There are also items — such as onion rings, desserts, and beverages — that have absolutely no impact on the sale of other products and are neutral because their demand is not linked to the demand for any other product. However, they remain on the menu because they add more to revenue than to cost.

In all probability food items that are substitutes for some are complementary for others. Those sales linkages will occur because menus offer more than one choice of item within each of the different categories, and the offerings in those categories compete with one another. Whenever a successful effort is made to increase the number sold of a particular menu item, it will reduce the sale of some other item. On the other hand, adding new items or reducing prices to encourage purchases will enhance revenue from customers who are drawn to the restaurant for the specials, but who instead opt to purchase regular menu items.

Armed with that knowledge, operators should set prices to enhance the sale of highly profitable items and reduce the sales of relatively unprofitable items in the same menu categories. The customer sensitivity to price differentials is even more critical in the purchase decision when the items are seen as identical substitutes. For example, Prime rib and sirloin steak are commonly offered in different portion sizes. They may be called "king" and "queen" cuts — the former being 14 ounces and the latter 10 ounces. But the only discernible difference is 4 ounces of meat. Customers quickly can calculate the cost per ounce, and the cost per ounce of the larger cut should be less than the smaller cut.

'Price-value' and 'value-pricing' are not one and the same

MANY OPERATORS, especially those in the fast-food sector, are resorting to what they call a "value-pricing" strategy. That strategy is a response to the successful campaigns launched by such competitors as Taco Bell and Wendy's that feature 59-cent and 99-cent menu items. McDonald's, in turn, has responded with its value-meal menu, which is the bundling of a sandwich, fries, and drink offered at a discounted price. Wendy's employs that pricing strategy in addition to its 99-cent menu choices. Value meals are simply a form of discounting the prices of regular menu items without the need for a coupon. The chains accept a lower margin and incur a higher food cost by selling at a lower price, thus the term value-pricing.

Contrast that to what Taco Bell has done with its menu. The chain's executives, who were seeking specifically to build customer traffic, realized that price is an important element in the purchase decision. But they did not simply cut prices on their existing menu. Instead, after taking a survey of its customers, Taco Bell developed a menu geared to the price points customers said they wanted to pay at lunch. That strategy marks a completely different approach to value-pricing and the traditional discount pricing.

We have seen that discounting is so pervasive in certain segments — particularly in the pizza sector — that regular prices rarely are charged. What Taco Bell did is really what is called "price-point pricing." Executives determined the specific price point that the chain's customers wanted to spend and then developed special menu items that could be sold at those prices and still return a decent gross profit and food cost.

Operators who attempt to compete by lowering prices of existing menu items will have to accept lower margins and higher costs. They cannot sustain that response for more than a short period of time, while Taco Bell can offer its items as part of the regular menu for as long as it chooses. Wendy's has accomplished much the same thing with its 99-cent menu. The difference, then, between cost-driven pricing and value-pricing is that the former starts with a competitive price and develops menu items to sell at that price, while the latter simply discounts menu items whose prices are cost-driven.

Know the competitive factors that influence the prices you can charge and the price points your customers will accept

WHILE THE CURRENT APPROACH to "value-pricing" is relevant, it only examines customer price sensitivity in the purchase decision. Price sensitivity is influenced by many other factors that must be considered when you are establishing menu prices. Certain factors allow you to charge more, while others require you to charge less. Certainly, the type of service delivery influences prices. If tableservice is provided, customers expect to pay more than if an operation is self-service. The location of a restaurant also will require and allow you to adjust prices accordingly. Consider the prices charged in resorts area versus those charged for comparable products and services in a residential area. We often use the term "tourist trap" to describe a place that charges far more than we prefer to spend. But this is a fact of life. We pay more for hot dogs and soft drinks at sports stadiums, convention centers, and theme parks.

In addition, elements that are perceived by the consumer as extras — such as amenities, ambiance, and entertainment — and are not offered at similar operations, can soften price sensitivity and influence the purchase decision. Having the lowest prices will not guarantee traffic if other aspects of the meal experience detract from the bargain a customer thought he was going to receive. The bitterness of poor quality lingers long after the sweetness of a cheap price is forgotten.

Remember: When certain "indirect cost factors" are present in your establishment, you can charge more the going market price (See key 5). By the same token, the absence of those factors may prevent you from setting prices on the high side of the pricing continuum. Those aspects exemplify why the pricing decision cannot be reduced to a simple markup of cost. You must incorporate a combination of subjective factors that involve a good amount of trial-and-error before you determine the right price for both you and the customer.

KEY

The average price of an entrée will influence your pricing of add-on items

IF YOU WANT TO SELL MORE appetizers, side orders, desserts, and bottles of wine, you should price them in relation to your average entrée prices. If the average price of your entrées is $12.95, the majority of your appetizers, side orders, and desserts should be priced between $3.25 and $6.50 — or 25 percent to 50 percent of the average entrée price. That price spread makes the à la carte items more appealing price-wise and therefore more likely to be purchased. Consequently, you must develop appetizers, side orders, and desserts within those price points that will allow you to meet your cost objectives.

The same thinking applies to pricing wines by the bottle, carafe, and glass. If the average check is $12.95, bottled wines should be priced between $16.25 and $24.95, with at least one

> **"In business, the competition will bite you if you keep running; if you stand still, they will swallow you."**
>
> — WILLIAM KNUDSEN

choice priced equal to your average check per customer. Wine-by-the-glass should be priced between 25 percent and 50 percent of your average per-person check. Once again, you must start with the price points and then select wines that will generate a reasonable profit. But be careful when pricing popular brands of wine that can be purchased at the supermarket and package stores. You won't make many sales at $24.00 for a bottle of Chardonnay that can be purchased for $6.00 retail.

Beer and wine — the beverages commonly consumed with food — can't sustain the same markup as mixed drinks. Beer and wine are marked up two or three times their cost, resulting in a 33-percent-to-50-percent cost-to-price. Microbrews and imports are extremely popular today, and pricing is demand-driven. Prices of $3.95 and up are not considered unreasonable for trendy beers. This is yet one more demonstration that pricing beverages and food are not simply a cost markup exercise. The subjective and indirect cost factors will temper the price upward or downward.

In addition the price spread between the highest- and lowest-priced item in any category shouldn't exceed two-and-a-half times the lowest-priced item in each respective category. That rule doesn't mean that you can't have items priced above that amount, but rather that 80 percent of the items should be within that range. It applies to pricing appetizers, side dishes, and desserts as well as entrées. For example, if you have five different appetizers, and the lowest is priced at $3.25, the highest should not exceed $8.25. Rather than reduce the price at the high end, raise the price at the lower end to $3.95 or $4.25 in order to bring the prices within range. Then the difference between the high and low is not so noticeable.

(1) Usually beer and wine can sustain the same markup as mixed drinks.

    A. True
    B. False

(2) Wine by the glass should be priced between:

    A. 40 and 60 percent of the average per-person check
    B. 70 and 80 percent of the average per-person check
    C. 25 and 50 percent of the average per-person check

(3) The bitterness of poor quality lingers long after the sweetness of a cheap price is forgotten.

    A. This statement is often true
    B. This statement is often false
    C. This statement is irrelevant to foodservice operations

(4) Usually if the average check is $12.95, most bottled wines should be priced:

    A. Between $24.95 and $29.95
    B. Between $21.95 and $24.95
    C. Between $16.25 and $24.95

(5) If you have five different appetizers and the lowest is priced at $3.25, the highest should not exceed:

    A. $8.25
    B. $5.95
    C. $6.75

ANSWERS: 1:B, 2:C, 3:A, 4:C, 5:A.

The perception of value to a buyer is a function of three elements

THE PERCEPTION OF VALUE is a function of three elements: (1) Quality; (2) Quantity; and (3) Price. Value to the customer is enhanced by an increase in the quality of the food or beverage; the product presentation; the service; ambiance and decor; the size of the portions; and the price. Value perceptions are relative. A higher-priced meal can constitute a value just as much as a lower-priced meal, depending on whether it is an eat-out or dine-out occasion.

A decrease in regular prices is often perceived as value, thus the attraction of discount promotions. However, we are seeing strong signs that today's customers are perfectly willing to pay the asking price for highly specialized products and services. Therefore, menu items that are uniquely prepared can be priced closer to the higher end of the continuum. Giving the

"I skate to where the puck is going to be, not where it is."

— WAYNE GRETZKY

---

customer the right combination of product quality, quantity, service, and price is the best strategy for building and maintaining market share. Those operators who can execute that strategy will find that product differentiation and branded products provide a competitive edge that can give them a distinct advantage in the marketplace.

Studies suggest that value-conscious customers will choose a dish priced somewhere between the highest- and lowest-priced items in a menu category. If that's true, you can use that knowledge to enhance the sale of certain items and, at the same time, please the customer. For example, if most of your entrées are priced in the $30-to-$40 range, try introducing a choice in the $20 range. A value-conscious customer is not likely to select a higher-priced item because it is perceived as too expensive, nor will she pick a $20 item because low price may be equated with a less-acceptable alternative. Therefore, the customer is likely to select something in the middle, or the $30 range. That strategy works particularly well for pricing wine by the bottle. If most of your bottles are priced in the $20-to-$30 range, try introducing one in the

$10 range. The extremes will not be selected, and it makes the $20 bottle the best value.

The value-to-price relationship is important because if the perceived value is seen to exceed the price, customer satisfaction will be achieved. Where spending is concerned, the key to customer satisfaction is always value and never price. You can take either a product approach or a customer approach with your pricing philosophy. The former concentrates on the food and service and emphasizes quality. Prices are set nearer the high end of the pricing continuum. If you take the latter approach, the needs of the customer will drive the product and the services you offer. It provides value and seeks repeat visits to build sales. Prices are competitive and moderate.

Those concepts of product- and customer-driven pricing are similar to the demand- and market-driven pricing discussed earlier. When those are combined with the psychological pricing theories of mental accounting, reference pricing, and odd-cents pricing, one begins to see how these intervening variables complicate the pricing decision.

(1) The perception of value is a function of which three elements:

    A. Quality, quantity, and ambiance
    B. Quality, price, and menu selection
    C. Price, quantity, and quality

(2) A decrease in regular price is often perceived as:

    A. A decrease in quality
    B. A decrease in quantity
    C. A discount value

(3) Studies suggest a value-conscious customer will choose a dish priced somewhere between the highest- and lowest-priced items in a menu category.

    A. True
    B. False

(4) Where spending is concerned, the key to customer satisfaction is always _____ and never price.

    A. Quantity
    B. Value
    C. Quality

(5) A higher-priced meal can never constitute a value just as much as a lower-priced meal.

    A. True
    B. False

ANSWERS: 1:C, 2:C, 3:A, 4:B, 5:B

Menu pricing is a method of managing revenue

IT IS GENERALLY ACCEPTED that the setting of prices is an important element of one's marketing strategy. We know that demand for certain items will change as the price is altered, thereby modifying the amount of revenue received. Getting the price right is one of the most fundamental and important management functions. Pricing your menu items to achieve a targeted minimum check average will enable you to achieve your revenue and profit objectives. In that context, the pricing decision becomes a critical element in the management of revenue.

The traditional cost-driven approach to pricing is no longer effective when it comes to managing revenue. Management-theory icon Peter Drucker said: "The third deadly sin (of business practice) is *cost-driven pricing*. The only thing that works is *price-driven costing*. . . . The only

> **" The most important of my discoveries has been suggested to me by my failures. "**

> — Sir Humphrey Davy

---

sound way to price is to start out with what the market is willing to pay — and designing to that price specification."

Menu prices will determine the average check and the total amount of revenue received from a day's transactions. With the growing number of restaurant choices, and all with comparable menu items, customer responses to price changes must be understood by the managers who make the pricing decisions.

If a price increase for a particular menu item results in fewer orders being sold, and the overall sales generated by that item is less than it was at the lower price, the price increase has actually hurt your revenue. However, if a menu item is increased in price, the number of orders sold does not decrease, and a gain in overall revenue is realized, the price increase was appropriate. That occurs with specialty items on which one has a "monopoly." It enables a restaurant to command a higher-than-average price, and hence, generate additional profits.

Depending on whether your restaurant is a casual-theme, fine-dining, quick-service, or family/coffee-shop operation, your flexibility in setting prices will differ. When you are selling "commodities" — menu items that are available at any number of competitors, such as hamburgers, chicken, and pizza — you will have less latitude in setting prices than you would if you were selling "specialty goods" — menu items that are unique to your operation and popular with the public, such as homemade entrées, appetizers, and desserts.

Specialty items return a higher profit margin than commodities. The important elements in pricing are demand and substitutions. Items in high demand that have no substitutes can command a higher price than those common items found on most of your competitors' menus. Therefore, without measures of customer demand and the behavior of your competitors, effective pricing decisions cannot be made.

The pricing objective is to identify the price that will increase customer counts, the average check, and gross profit, while, at the same time, optimizing overall sales revenue relative to achieving the lowest overall food-cost percentage. Remember: Low prices put tremendous pressures on operating margins. Given the choice, take higher margins over lower margins every time.

(1) The traditional cost-driven approach is still the most effective revenue management tool.

    A. True
    B. False

(2) A "specialty item" might be:

    A. Hamburgers
    B. Pizza
    C. Fresh pasta

(3) Low prices put tremendous pressure on:

    A. Customers' menu decisions
    B. Operating margins
    C. Labor costs

(4) Given the choice, take:

    A. Higher margins over lower margins every time
    B. Higher margins over lower margins sometimes
    C. Higher margins over lower margins two out of three times

(5) Specialty items should never return a higher profit margin than commodities.

    A. True
    B. False

ANSWERS: 1: B, 2:C, 3:B, 4:A, 5:B.

How to cost and price all-you-can-eat salad and food bars

THE MOST WIDESPREAD PRICING method for salad and hot-food bars is one price for all-you-can-eat. Pricing food prepared and plated in the kitchen is somewhat easier than pricing salad bars and buffets where portioning is not controlled and the waste factors are greater. However, you can cost out the average portion to get an idea on the food cost for each customer served.

The per-person pricing decision begins by determining the food cost for all of the items required to set up the bar, including the backup inventory used to replenish it during the meal period. Assume that $1,000.00 represents the beginning inventory of food at the start of the day, and that $350.00 represents the value of the food that was left over at the end of the day. The cost of food consumed is the difference between the opening and ending inventory, or $650.00. If the number of customers going through the buffet that day was 225, the average food cost per customer was $2.89 ($650.00/225).

That procedure should be repeated for each meal period for two or three weeks. The average food cost per customer will vary slightly from day to day, based on the customer mix, meal period, day of the week, and items offered on the buffet (if the menu changes). A weighted average can be calculated to approximate the standard cost per customer and the price based on that standard portion. This method averages the men, women, teenagers, and young and old customers who pass through the buffet. The average cost per person serves as the starting point for your pricing decision. The price will reflect both objective and subjective factors and direct and indirect costs (See keys 1-9).

With one price for an all-you-can-eat buffet, the light eaters cancel out the heavy eaters because they pay as much as the customers who return a second and third time. Operators seek to avoid the challenges that undoubtedly occur when there are multiple price categories for children, adults, and seniors. The one price for adults and one for children under 12 usually are the only price categories.

How to cost and price by-the-ounce salad and food bars

RESTAURANTS AND INSTITUTIONAL foodservice operations are experimenting with pricing by-the-ounce as an alternative to the price per person for all-you-can-eat. One of the ideas they're testing is whether pricing by-the-ounce will increase the marketing value of all-you-can-eat promotions to nutrition- and diet-conscious patrons. In addition, all-you-can-eat operators are tiring of having to police customers from taking too much, asking for doggie bags, and sharing with nonpaying members of their parties. In addition, by-the-ounce lends itself better to carry-out than the one price for all-you-can-eat. Restaurants can charge customers for what they take, and customers can pay for only what they take.

Arriving at a by-the-ounce price is complex because the food cost of items offered on a

> **"Good habits are as easy to form as bad ones."**

> — TIM MCCARVER

food bar ranges from inexpensive croutons and bean sprouts to the more costly chicken, shrimp, and salmon salad. The per-person tabs for all-you-can-eat range from a low of $1.99 for a basic salad bar to double-digit checks for elaborate buffets that offer meat and seafood selections. How does one arrive at a price that is acceptable to the customer's price-value perceptions and the operator's profit and food-cost goals?

The process of arriving at an average cost per ounce of the ingredients on the food and salad bar is a little more complicated than determining the average food cost per person for the food bars. The process must account for the extreme differences in density and volume of one ounce of bean sprouts and one ounce of tuna or shrimp salad. For example, three ounces of bean sprouts will fill a medium-sized salad bowl, while three ounces of potato salad would amount to only three or four tablespoons. Items usually measured in liquid ounces must now be measured in avoirdupois ounces.

When costing by-the-ounce, all items must be weighed. Divide the weight, expressed in ounces, into the total recipe cost to arrive at the cost per ounce for each item. Calculations of some common food and salad bar items revealed costs per ounce ranging from two cents to 30 cents. If a cost markup is based on the highest cost item, the price per ounce won't communicate price-value to the customer. When pricing by-the-ounce, consider the customer's perspective concerning quality, quantity, the variety of items offered, and the type of operation. For example, students and faculty patronizing a college cafeteria expect to find lower prices than if they were at a commercial cafeteria in a shopping mall. There are definite price points for each concept, menu, and service-delivery system that must be considered before setting the ultimate price.

Establishing the price per ounce that will achieve a specific average check is important. First, estimate the weight of an average portion and divide the number of ounces into the desired average check amount. If you need to average $3.95 at lunch, and the average portion size is 16 ounces, a price of 25 cents per ounce is necessary.

(1) When costing by the ounce, all items must be:

    A. Frozen
    B. Weighed
    C. Counted

(2) With one price for an all-you-can-eat buffet, the light eaters rarely cancel out the heavy eaters.

    A. True
    B. False

(3) In figuring by-the-ounce pricing, you must account for the difference between:

    A. The volume and density of different items
    B. The ambiance and service of the restaurant
    C. The texture and density of different items

(4) The two pricing alternatives for salad bars and buffets can be described as:

    A. On or off the menu
    B. Cost-reflective assumption marketing
    C. By-the-ounce or per-person

(5) There are definite price points to be considered for each concept, menu, and service-delivery system that must be considered before setting the ultimate price.

    A. True
    B. False

ANSWERS: 1:B, 2:B, 3:A, 4:C, 5:A.

Strategies for taking the anxiety out of increasing menu price

RAISING MENU PRICES is one of the most anxiety-provoking tasks a restaurateur has to shoulder. How to go about it, when to do it, and how much to raise the price are all important considerations in the decision-making process. The anxiety is warranted too, because customers often respond adversely whenever they discover that prices have been hiked. The feedback is expressed not only in the form of verbal comments but also in declining customer counts.

Consequently, the task of raising prices is beset with misgiving and uncertainty. Anxiety results from fears that the approach to pricing is not the best or correct one, given the respective menu item, competitive conditions, and customer perceptions. Regardless, you must adjust prices to maintain your profit objectives, and

> **"A mistake at least proves that somebody stopped talking long enough to DO something."**
>
> — ANONYMOUS

accomplish it in a manner that doesn't drive away your regular customers.

Increases in menu prices must be made as subtly and unobtrusively as possible. The less attention you call to price increases, the less chance there will be for adverse consequences. The following suggestions are offered as ways to downplay and soften the negative customer reaction that sometimes follow price increases.

(1) Use such odd-cents increments as 25 cents, 50 cents, 75 cents, and 95 cents for digits to the right of the decimal point. An item increased from $7.75 to $7.95 is less likely to be noticed by the customer. Consequently, you give yourself three opportunities to increase your prices before the dollar digit must be changed. Some operators price their menus at the exact penny — $4.72 or $4.86. But the two most popular terminal digits are "5" and "9" — for example, $4.95 or $4.99. The customer interprets these prices as "four dollars and change," which is psychologically cheaper than "five dollars."

(2) Never raise prices when you print a new menu, especially if you change the design and format. Regular customers are more likely to notice the addition or deletion of menu items along with price increases. If you're transitioning to a new menu format and need to raise prices on popular menu items, make the price changes on the last reprint of the old menu so you won't have to raise prices on the new one. You will be able to point out that the new menu offers the old menu prices.

(3) It's not worth the savings on printing simply to cross out or place tabs over old prices. You're really calling attention to the increases, which will generate questions from customers. You're better off reprinting the menu with the changes. However, if you want to reduce prices for a special promotion, crossing out the higher price and writing in a lower one works the same way it does in a department store when a sale is being held. It really projects a bargain in the mind of the customer.

(4) The one price hike that is most likely to be noticed is when you have to increase the dollar amount — going from $9.95 to $10.25, for instance. Hold off as long as you can with that type of increase by reducing portions or accompaniments instead of raising the price. When you must increase it, try repositioning the item in a less-noticeable location. If that is not practical because it is a signature item, increase the portion size or add an accompaniment to create a new and improved menu item.

(5) Whenever portions and accompaniments are already substantial, consider reducing portion sizes or eliminating one of the

accompaniments to reduce your cost. That accomplishes the same task as raising the price and isn't as noticeable. Many operations have dropped either the salad or the potato on lunches and dinners instead of increasing prices. That may be the appropriate strategy in highly competitive markets.

Back in the mid-1970s, my restaurant opened with a menu that offered a free glass of beer or wine with every entrée. The cost of the beer and wine was factored into the food ingredients in the price of the menu item. After two menu-price increases in three years, we noticed our customer counts dropping. As a result, we grew reluctant to raise prices, although our overhead and product costs were increasing. Finally, we made the decision to eliminate the free beer and wine and placed a sticker on the menu stating, "In lieu of increasing our prices to cover the increasing costs of doing business in these inflationary times, we have eliminated the free glass of beer or wine. We will continue to use only top-quality ingredients in all our menu items and on our antipasto salad bar."

(6) Never raise prices across the board. It's rarely warranted and definitely will be noticed by your clientele, which will result in declining customer counts. I recall a restaurant that did that after about six months in operation. One day we went in, and every item on the menu had been increased by one dollar. Two months later that restaurant was out of business. The amount of the increase and the fact that it impacted every entrée was not the right decision. You're better off raising a few items at a time, beginning with the most popular menu

items. A small, incremental price increase — 25 cents, for example — on a popular item will be less noticeable and generate more revenue than if you had increased a slow-selling item by a noticeable amount.

(7) Items that fluctuate in cost on a weekly or monthly basis — fresh fish and seafood, for instance — shouldn't be priced at all. The menu should simply list these items as "market priced," and the server should quote a price on a daily basis.

(8) Avoid a common menu-design technique of aligning prices in a straight line down the right side of a column or page. You've probably seen the menus with the line of dots running from the last word in the menu description to the extreme right-hand margin. That format makes the prices stand out from the menu copy. Customers will look at the price, and then read the copy, basing their choice more on price than on the item itself. To make prices less prominent, place the price immediately after the last word in the menu description. Furthermore, never place items in a descending order that lists prices from high to low, or vice versa. Mix them up. And, of course, keep in mind which items are your newest and most popular when determining menu order.

DAVE PAVESIC is a former restaurateur who now teaches hospitality administration at the university level. He previously owned and operated two casual-theme Italian restaurants in Orlando, Fla.; served as general manager of operations of a six-unit regional chain in the Midwest, operating four coffee shops, a fine-dining seafood restaurant and one drive-in; and was a college foodservice director. He currently teaches courses on restaurant cost control, financial management, and food production in the Cecil B. Day School of Hospitality Administration at Georgia State University in Atlanta, Ga. He has written numerous articles on menu-sales analysis, labor cost, menu pricing and equipment layout. His two other books are *The Fundamental Principles of Restaurant Cost Control*, Prentice Hall Publishers, 1998, ISBN 0-13-747999-9 and *Menu Pricing and Strategy*, fourth edition, Van Nostrand Reinhold Publishers, 1996, ISBN 0-471-28747-4.